A Simple Guide to Popular Physics

An introduction to particles, quantum physics and cosmology for absolute beginners.

Tony Harris

First published 2020 by RAH publishing

ISBN 978-1-8380697-5-9

For my young grandchildren, Jake and Ewan, who will learn much more than I'll ever know.

Acknowledgements

Thanks to Joy for her help and encouragement.

Thanks to Alison Jack my amazing editor. (alisoneditor@outlook.com).

Author's note

Thank you for buying my book.

When I first decided to write and publish a book on popular physics, I did not realize how much I would enjoy the experience. My hope is that you will enjoy reading it as much as I enjoyed writing it.

My plan was to write a book that I wish I had read before embarking on my years of reading popular physics. A book to give readers a good, clear and simple grounding in this fascinating genre.

How well I have succeeded in this ambition can only be decided by you, my readers. If you enjoy this book, or have any comments, I would be most grateful if you would leave a review on the Amazon website or elsewhere.

Tony Harris is a retired teacher living in the UK.

"There comes a point in your life when you need to stop reading other people's books and write your own."

Albert Einstein

"If you can't explain it simply, you don't understand it well enough."

Albert Einstein

Contents

Introduction

This book offers the casual reader an insight into the physics of nature. It is by no means a comprehensive study; rather, it assumes that the reader has little or no previous knowledge of the subject. Anyone from teens to centenarians can derive a basic grounding in particle physics, quantum physics and cosmology from this book. And the really good news is that there are no mathematics or mathematical equations involved, apart from one that I'm sure you've all heard before:

$$E = mc^2$$

On a minute scale, we have particle physics and quantum mechanics, and on a huge scale, we have the cosmos. None of these subjects can be considered "easy", but my aim is to offer you as simple an understanding as possible. That is why I will explain every theory and concept I cover in this book concisely to give you a starting point. A basic understanding of these concepts and theories will lead you to a better understanding of the more detailed popular science books. After all, it is good advice to learn to walk before you run.

Chapter 1 – The Professor's Diamond

Figure 1 – The Professor's Diamond[1]

In 2011, the British Broadcasting Corporation (BBC) broadcast a lecture called *A Night with the Stars*, presented by Professor Brian Cox. The studio audience was made up of well-known British celebrities and there were millions of viewers.

The subject was quantum mechanics.

---

[1] Professor's Diamond, Lifetime stock/Shutterstock

The entertaining and informative broadcast was on the whole well received, but there were some viewers who took exception to an assertion by Professor Cox towards the end of his lecture. A few of these were physicists themselves.

Professor Cox had a prop to help with his explanation of Pauli's exclusion principle, which says:

No two electrons can be in the same quantum state in atoms.

We will explore Pauli's Principle in more detail later in the book. Suffice to say now that Professor Cox gave a good summary of the exclusion principle, and then produced a large diamond. He rubbed the diamond between his hands and informed the audience that this action had altered the energy levels of not only the three million, million, billion electrons in the diamond, but all the electrons in the universe.

I admit that I was rather surprised and more than a little sceptical at this statement. As a result, I spent a considerable amount of time researching the professor's statement and revising my previous reading. By the end of my research, I realised I had increased my knowledge substantially by questioning Professor Cox's simple assertion.

In making this statement in his popular lecture, broadcast on prime-time television, Professor Cox had shown that in quantum physics, everything is connected. His critics had missed the point; the professor knew his audience. If he had gone into detail about the exclusion principle and the complicated concepts and mathematics involved, he would have lost their attention.

He was also limited by the airtime of the programme. To explain that no two electrons can be in the same quantum state in a few minutes is impossible. Instead, he used just one parameter of the four required to describe an electron's quantum state: energy levels. Energy levels in Pauli's exclusion principle are relatively easy to explain. In doing so, Professor Cox simplified the concept and put his point across to people without a PhD in Physics.

Brian Cox is not only a reputable physicist; as far as laypeople who take an interest in popular science are concerned, he is a brilliant communicator. Those of us who are not reputable physicists can still enjoy popular science as informative and entertaining if we accept our limitations.

The value of questioning

If you notice inconsistencies or ambiguities in this book please try to resolve them yourself . In particular you may discover an ambiguity regarding the speed of light. If you do try to resolve it. There are many books and resources available that will take you into more detail; my aim is simply to whet your appetite. Quite simply, this book is not an academic enterprise. I have written it to encourage you to look deeper into the thought-provoking and enjoyable genre of Popular physics.

Chapter 2 – Some Basics of Classical Physics

Before we have a look at particle physics, quantum theory and the cosmos, we need a brief introduction to the concepts of classical physics:

- Energy
- Weight and mass
- Matter – solids and liquids
- Measures and units

Energy

Energy must be transferred to an object in order to perform work on or heat it.

Newton's law of the conservation of energy states that it may be transformed from one form to another. *It cannot be created or destroyed.* The SI unit for energy is the joule (SI units are explained briefly later in this chapter). Mass and energy are closely related, as we will discover in Chapter 4, when we discuss Einstein's theory of Special Relativity.

FORMS OF ENERGY

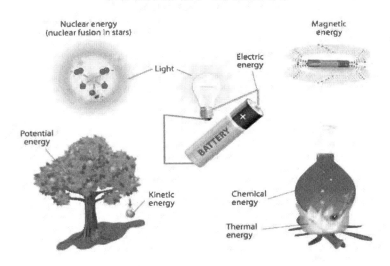

Figure 2[2]

Types of energy include:

- Kinetic energy (movement)

- Chemical energy (e.g. coal, natural gas, etc.)

- Thermal energy

- Magnetic energy

- Light energy

- Electric energy

2 **Designua/Shutterstock**

- Gravitational potential energy
- Nuclear energy

Weight and mass

- **Weight** changes when the pull of gravity changes. On Earth, an object will weigh six times more than it weighs on the Moon. Earth, being larger than the Moon, has a larger gravitational pull. Weight is therefore dependent on the gravitational force where the object is situated. According to the International System of Units (SI Units), weight is measured in Newtons with the symbol N. There is more information on SI units later in this chapter.

- **Mass** can be described simply as the amount of matter there is in an object. This measurement is given in kilograms or grams and is calculated by multiplying the object's volume by its density (see below). An object's mass will be the same wherever it is measured because the object will always contain the same number of protons (amount of matter).

- **Matter.** There are several forms of matter that we know of, but only three are of relevance to this book: solids, liquids and gases.

Matter

Solids, liquids and gases:

- **Solids.** Solid objects have a defined shape because their atoms are packed tightly together (i.e. they have a high density). The atoms cannot move around and cannot be compressed into a smaller volume.
- **Liquids.** In liquids, the atoms are not so tightly packed so they can flow around each other. Most liquids can be compressed into a smaller volume in a container, i.e. their atoms are forced closer together (they become denser).
- **Gases.** The atoms in gases are in constant movement and have a relatively large space between them. Gases can be compressed into a smaller volume when confined in a container and expand when released.

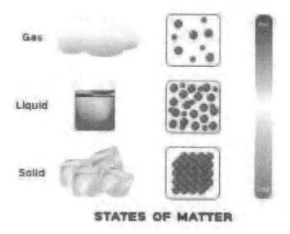

STATES OF MATTER

Figure 3[3]

Measures and units

- **Density.** The density of matter is a measure of how closely the atoms are packed together. Density is measured in kilograms per cubic metre and can be calculated by dividing an object's mass by its volume.

- **Volume.** The amount of space that matter occupies. There are many ways of measuring volume depending on whether you are measuring solids, liquids or gases. The formulae for measuring different

---

[3] MicroOne/Shutterstock

shapes of solids or containers of liquids or gases are different.

As I promised in the introduction not to burden you with mathematics, other than $E=mc^2$, I will leave you to find these formulae for yourself in further reading if you are interested. There are, however, some easy ways to measure small amounts of gases and liquids (see Fig 4 and 5). Figure 4 shows gas volume being measured by water displacement. Figure 5 shows measuring vessels for liquids.

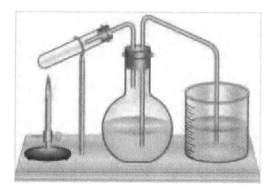

Figure 4[4]

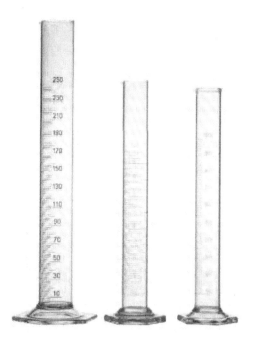

Figure 5[5]

[4] Fouad A Saad/Shutterstock

[5] PRILL/Shutterstock

The units in which volume is measured can be confusing. The customary system in the US differs from the imperial system in the UK, and both of those differ from the metric system. In 1960, an international system of units was introduced. Le Système Internationale d'Unités, now known simply as SI Units, is used by the scientific community in order to avoid confusion.

There are seven base units:

- The kilogram (kg) for mass
- The second (s) for time
- The kelvin (k) for temperature
- The ampere (a) for electric current
- The mole (mol) for the amount of a substance
- The candela (cd) for luminous intensity
- The metre (m) for distance

There are many other units derived from these seven. If you are interested, you can easily find these online.

Chapter 3 – What is Reality?

All creatures on Earth have evolved in ways which give them an advantage in their own environment.

In this chapter, we will cover:

- The reality of our senses
- Technology
- The reality of the tiny and the huge

The reality of our senses

Everything we perceive or know is constrained by the limitations of the human brain and our five senses.

All species of life on our planet see reality in different ways. A fish swimming in an ocean will have no idea of anything outside its own environment. A wolf views reality in a different way to a human or a bird because its senses have evolved in a different way. All creatures on Earth have evolved in ways which give them an advantage in their own environment.

To paraphrase the writer Anais Nin: 'We don't see things as they are, we see things as we see them'.

From this perspective, we can see that there are millions of different realities on just one planet. Humans are set apart from all other creatures on our planet by our consciousness, our bigger brains and ability to acquire knowledge and understand complex concepts. Because of the knowledge we have acquired, we know we cannot rely on our senses alone to discover reality. Our notion of true reality is far from complete. We are still at the beginning of our understanding of the world around us, let alone the universe. Like the fish we have no notion of what is beyond the boundaries of our knowledge but unlike the fish, we know there is something.

What is dark energy? What is dark matter? What happened in the milliseconds after (or before) the big bang? Are there smaller particles than those we've already discovered? These are some of the many mysteries yet to be resolved. The human race will no doubt introduce new technologies in the future that will assist in expanding our knowledge of the universe, but there is a long way to go before we can be sure that our notion of reality is correct.

Most scientists now agree that, physically, the human brain has hardly altered since the stone age, but our perception of reality has changed over the course of human history. It is our acquisition of knowledge and the subsequent growth of technology that has developed over this period of time. And with the rate of technological change now increasing rapidly, our acquisition of knowledge is also increasing faster than ever.

The 20th century gave us Einstein's special relativity (in 1905) and general relativity (in 1915). We will cover these in more detail in Chapter 4. It also gave us quantum theory (also known as quantum mechanics), which we will look at in Chapter 6. These are the most successful theories to date. They have been tested experimentally and mathematically many times and have been found to be correct.

Technology

Technology has given us the Hubble Space Telescope, launched in1990 and named after Edwin Hubble, of whom we will read more in Chapter 7. A reflecting type of telescope with a 94.5 inch (2.4 metre) mirror, it has produced fabulous images from up to 15 billion light years away and has enhanced astronomers' work hugely.

The Large Hadron Collider (LHC), constructed near Geneva, Switzerland at the European Organisation for Nuclear Research (CERN), first became operational in 2008. It is designed to propel subatomic particles around its 27 km tube at very close to the speed of light, and then collide them together with tremendous force which smashes them into tiny parts. Its purpose is to test the predictions of physics theories and search for any undiscovered particles which could extend our knowledge of physics and reality.

The Scanning Transmission Electron Holography Microscope (STEHM), the world's most powerful microscope, became operational in 2013 at the University of Victoria in Canada. It has enabled physicists to see a much higher resolution image of the atom than ever before. As microscopes go, it is massive, weighing 7 tonnes and standing 4.5 metres high.

The reality of the tiny and the huge

"I think I can safely say that nobody understands quantum mechanics."

Richard Feynman, physicist

Quantum theory can seem counterintuitive. In fact, it doesn't even make complete sense to those who contributed to its formulation.

Particle physics (the properties, relationships and interactions of subatomic – smaller than an atom – particles), quantum mechanics (the mathematical description of the motion and interaction of subatomic particles) and cosmology (the science of the origin and development of the universe) are intrinsically intertwined. Thanks to these theories, physicists have brought us closer to true reality than ever before. They have taken us back to within milliseconds of the Big Bang, giving us knowledge of subatomic particles and the vastness of the universe.

The physicists' holy grail is a "theory of everything", a grand unified theory, and it's now a serious possibility. The current challenge to developing a theory of everything is to match Einstein's gravity, from his 1915 theory of general relativity, and quantum theory. Unfortunately for most of us, a clear understanding of modern physics is not an option. The closest we can come is via popular science,

Chapter 4 – Einstein

This is not the place for an in-depth study of Albert Einstein's revolutionary theories of relativity. I'm simply going to outline for the purpose of this book:

- Special relativity
- General relativity

Special relativity

In 1905, Einstein published his theory of special relativity. Before this, scientists believed that time was absolute – it was the same everywhere and passed at the same rate throughout the universe. To them, time and space were unrelated concepts.

Prior to the 20th century, scientists believed that space was filled with a mysterious and highly elastic medium which they called the aether. They thought this substance allowed light to travel through the vacuum of empty space. They also believed that mass and energy were two distinct, unrelated concepts.

When Einstein published his theory of special relativity in 1905. He explained that time does not pass identically everywhere; it ticks differently wherever you are in the universe. Our current notion of time is a result of the effects of Einstein's theory of relativity. This doesn't affect us in our everyday lives because the time differences are very small on Earth.

In simple terms, imagine you are standing on a station platform and an express train passes, travelling at high speed. You observe a passenger sitting on the train. Time for that passenger is passing more slowly relative to the stationary you, but not enough for you to perceive.

At faster speeds and longer times, it would become more noticeable. Take an example of one of a pair of identical twins travelling at close to the speed of light to somewhere in Outer Space. On his return to Earth, he would find that his sibling had aged more than he had. His twin may even be long dead from old age, depending on the amount of Earth time that had passed.

Time passes more slowly at high speed and objects become distorted. Einstein proposed that time and space were interwoven into a single continuum, which was later described as spacetime. As the name suggests, spacetime mixes up time and space.

Einstein showed that light can travel through empty space without a medium and, in a separate 1905 paper, he collaborated with other physicists to show that light is both particle and wave (the duality of light is explained in Chapter 6). Light travels at a constant velocity in a vacuum, in every direction and to every observer, regardless of their position or speed.

Nothing can travel faster than light.

He also explained that mass and energy have equivalence. Mass can be turned into energy and energy can be turned, theoretically, into mass. Denoting energy as *e*, mass as *m* and the speed of light as *c*, he came up with the most famous equation in the world:

Energy equals mass multiplied by the speed of light squared.

$E=mc^2$

The speed of light in a vacuum is almost 300,000 kilometres per second (186,000 miles per second). The speed of light squared (300,000 x 300,000) = 90,000,000,000 (90 billion) kilometres per second. So $e = m$ x 90,000,000,000 km/s. It is a huge number.

This shows how much energy there can be in even a small mass of matter, which is the reason why tiny amounts of Uranium and Plutonium can cause massive atomic explosions and be used to fuel nuclear power stations.

Physicists have long acknowledged Einstein's premise that energy can be turned into matter by turning his equation around to $m = e/c^2$ (mass equals energy divided by the speed of light squared), but because the speed of light squared is such a huge number, it would take a huge amount of energy to result in a tiny amount of matter.

Special relativity applies within a frame of reference in which an object remains at rest or moves at constant speed. It does not take acceleration into account.

General relativity

Einstein spent ten years after publishing his theory of special relativity working on his theory of general relativity, which he published in 1915. This theory takes into account bodies which are accelerating and offers a new theory for gravity.

Before Einstein, Newton's theory of gravity, published in 1687, was universally accepted. Newton held that gravity is a force that works at a distance. The force pulling between two bodies depends on how massive the bodies are and the distance between them. Each body pulls from its centre towards the other body's centre. For example, the force of gravity from the Sun's centre pulls on the Earth's centre. The Earth, being less massive than the Sun, is held by the more massive body. Similarly, the reason we are held on the surface of Earth is that the planet is much more massive than we are.

However, gravity is a weak force. You can see for yourself how weak it is. Place a metal key on a table. Hover a small magnet over it. The key, which is held on the table by the gravity of the whole planet, will be lifted by the small amount of electromagnetic force in the magnet overcoming Earth's gravity.

Einstein, in his theory of general relativity, realised that gravity is not a force between masses but an effect of the warping of spacetime in the presence of matter (stars, planets, etc.). Space and time are interwoven into a single continuum. It is four dimensional (our normal three dimensions plus time).

It's difficult, if not impossible, for the human brain to visualise four dimensions, so the continuum is usually depicted as a flat rubber sheet with massive objects (stars, galaxies, planets) resting on it.

Figure 6[6]

Curved spacetime causes light to bend around massive objects. This is known as gravitational lensing and is used by scientists to see behind massive objects such as galaxies. The light from the hidden object bends around the obstructing galaxy to reach us on Earth.

Gravitational lensing is also used to see where dark matter is situated in the universe. We will be discussing dark matter in Chapter 7.

---

[6] Rosta9/Shutterstock

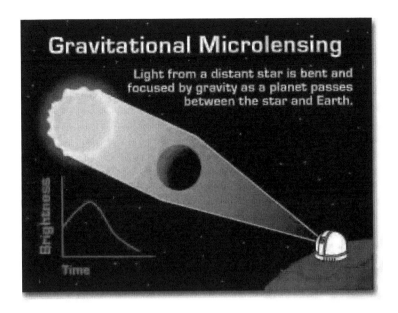

Figure 7[7]

One of the predictions of general relativity is that gravitational waves ripple through the universe. Gravitational waves are caused by massive bodies such as black holes and neutron stars orbiting each other and accelerating or colliding. The waves ripple out like water waves in a pond when a stone is thrown in. They cause the fabric of space time to distort as they travel through space at the speed of light, squeezing and stretching everything in their path and weakening as they go. The objects causing the waves can be billions of light years away from us.

---

[7] NASA/Wiki commons

The consequence of this is that they are very weak when they reach us. This explains why gravity is so weak here on Earth.

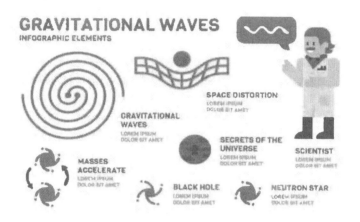

Figure 8[8]

Gravitational waves were first detected on Earth in 2016. They were produced by two black holes colliding and reached us after travelling at the speed of light for 1.3 billion years. Physicists are currently searching for a particle called a graviton which, theoretically, could be the carrier of gravitational waves. This would take them a step closer to a theory of quantum gravity and a grand unified theory, making general relativity and quantum theory compatible.

---

[8] **narakOrn/Shutterstock**

Chapter 5 – The Atom

In this chapter, we will cover:

- The structure of the atom
- Dalton's atomic model
- Thomson's atomic model
- Rutherford's atomic model
- Bohr's atomic model
- The neutron and proton
- The forces of nature
- The neutrino and electron neutrino
- Antimatter
- The Periodic Table of Elements

The structure of the atom

Physicists have deduced that a single atom of carbon has a diameter of 0.00000003 centimetres. As atoms are so tiny, scientists decided that it is more useful to compare atoms' size relative to each other, using carbon as the standard. The diameters of all other atoms can be three times bigger or three times smaller than that of an atom of carbon.

The human eye cannot see something so small. Until 2013, we couldn't even see atoms clearly through a microscope.

The particles inside the atom are very much smaller than the atom itself. Elementary particles cannot be divided into smaller parts as they have no internal parts. All elementary particles of the same type are identical and in perfect condition. Different atoms are only distinguishable by the number of electrons they have and how they are arranged. For example, hydrogen has one electron, oxygen has eight and iron has twenty-six.

Let's now look at developments which have led us to our current knowledge of particles, forces and atomic structure. There have been several attempts at explaining the structure of the atom (see Figure 9).

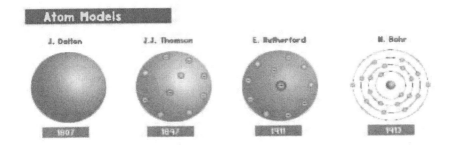

Figure 9[9]

Dalton's atomic model

In 1807, John Dalton proposed his atomic theory. He suggested that the atom was the smallest form of any element; it had no internal particles (see Fig 9). He believed that all atoms of a particular element were the same.

At this point in time, there was no evidence for any of the elementary particles which have since been discovered. Gravity had been discovered by Isaac Newton in 1687. The other three fundamental interactions (forces) – electromagnetic, strong nuclear and weak nuclear – were unknown in 1807. (We will discuss the forces of nature later in this chapter).

---

[9] Emir Kaan/Shutterstock

After work by scientists such as Hans Christian Ørsted in the early years of the 19th century, James Clerk Maxwell published his paper on the electromagnetic field in 1865. In 1897, Joseph John Thomson published a paper detailing the results of experiments using cathode rays, during which he'd found that these rays were not composed of light, but of negatively charged particles.

He had discovered the electron.

Until then, the smallest known piece of matter had been the hydrogen atom. The electron is 1,000 times smaller and 1,800 times lighter than the hydrogen atom.

Thomson's atomic model

Thomson's model of the atom was presented shortly after the electron's discovery. In this model, consisting of a sphere of positively charged matter containing negatively charged particles (electrons), his atom was neutral. Thomson was also the first scientist to suggest that atoms were divisible. His model is known as the plum pudding model because, well, it looks like a plum pudding.

At this time, the only known subatomic particle was the electron. Apart from gravity, the only known force of nature was the electromagnetic interaction (force).

Rutherford's atomic model

In 1911, Ernest Rutherford published his atomic model after he and his associates had conducted an experiment known as the gold foil experiment, for which they constructed an apparatus that fired alpha particles through gold foil (see Fig 10). Alpha particles are produced by the decay of a radioactive atom's nucleus. The atom's nucleus is unstable, so it ejects the alpha particle in order to make itself more stable.

Rutherford and his associates used gold foil because it can be stretched until it is very thin. They were able to make the foil just 1,000 atoms thick.

One of the properties of alpha particles is their inability to penetrate far into matter. The experiment showed that the particles could penetrate into the foil unless they hit an atom's nucleus, at which point they were deflected.

Rutherford's Gold Foil Experiment

Most α Particles Travel through
the Foil Undeflectd

Some α Particles are
Deflected by Small Angles

Gold
Foil

Few α Particles Travels
Back from the Foil

Detector

Beam of α Particles

Radioactive Source

Lead Shield

Figure 10[10]

10
udaix/Shutterstock

The result of this experiment inferred that atoms had a tiny, dense positively charged nucleus at their centre. The alpha particles that hit it bounced back. The particles that did not hit it went straight through.

An analogy to show just how tiny the atom's nucleus is may help here. It is easy to forget when looking at not-to-scale atomic diagrams. The LHC at CERN in Switzerland has a diameter of 8.5 kilometres (5.3 miles). If this was the diameter of an atom, the nucleus would be the size of a tennis ball.

Most of the atom's mass is contained in the tiny nucleus, so nearly all of the atom is empty space. This being the case, why do structures not collapse? Why do you not fall through your living room floor? I will explain further in Chapter 6 when we examine Pauli's exclusion principle.

In Rutherford's model, the number of negatively charged electrons around the nucleus is the same as the total charge in the atom's nucleus. This explained the neutrality of the atom, but the model could not explain how the negatively charged electrons did not radiate energy and fall into the positively charged nucleus. This problem makes the atoms in Rutherford's model unstable.

Now we had the electron, the electromagnetic force and a tiny, dense positively charged nucleus at the atom's centre.

Bohr's atomic model

In 1913, Niels Bohr introduced a theory of atomic structure based on quantum ideas. This model adapted and improved Rutherford's model and was an attempt to explain why the electrons did not radiate into the nucleus. Bohr's model again had a tiny positively charged nucleus at its centre and negatively charged electrons orbiting in the relatively huge space around it. But this time, the electrons were arranged in fixed spherical orbits or "shells" around the nucleus.

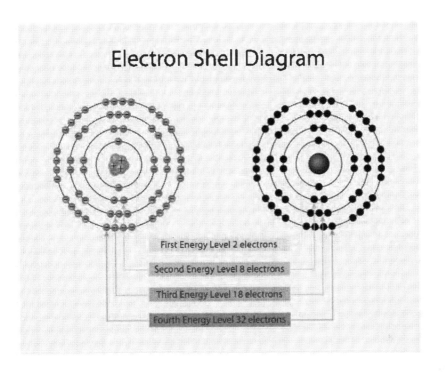

Electron Shell Diagram

First Energy Level 2 electrons
Second Energy Level 8 electrons
Third Energy Level 18 electrons
Fourth Energy Level 32 electrons

Figure 11[11]

The diagram on the right of Figure 11 shows the particles that Bohr would have known in 1913: the electrons and a dense nucleus. He would also have been familiar with the electromagnetic force. Bohr knew from the work Einstein did, expanding on the German physicist Max Planck's proposal, that electrons can emit photons (packets of energy) in fixed packages (quanta). Electrons are a type of

---

[11] Naski/Shutterstock

elementary particle called fermions. Photons are also elementary particles called bosons. We will look at both in more detail later in this chapter.

The atom fills up from the orbit nearest to the nucleus, which is known as the ground state. The orbits above ground state are known as excited states. Each orbit has a maximum number of electrons it can hold, which are shown in Figure 11, and each level out from ground state has an increased energy level. Electrons can drop a level by emitting a photon or jump up a level by gaining a photon (see Fig 12).

This solved the problem of Rutherford's model, which did not explain why negatively charged electrons didn't fall into the positively charged nucleus. Although this model has problems, it is still used in schools to explain Pauli's exclusion principle. We will touch upon this principle in Chapter 6, along with the modern electron cloud model.

One of the problems with Bohr's model, though Bohr could not have known it at the time, was that it would violate Heisenberg's Uncertainty Principle, introduced in 1926. There is more on this principle in Chapter 6.

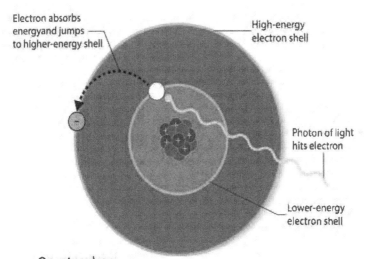

Electron absorbs energyand jumps to higher-energy shell

High-energy electron shell

Photon of light hits electron

Lower-energy electron shell

Quantum leap
Electrons in an atom can only jump directly
from one energy level, to another - a Quantum leap,
They cannot occupy an intermediate energy level.
When they move between levels,
electrons absorb or emit energy.

Figure 12[12]

The electrons are never in the space between the energy levels. They simply disappear from one and appear in another. These are the famous quantum leaps, which are just one example of the many counterintuitive actions in quantum mechanics.

12 Naski/Shutterstock

The neutron and proton

The diagram on the left of Figure 11 (the electron shell diagram) shows the nucleus containing two new particles which were unknown to Bohr in 1913. These are the proton and the neutron. The proton is positively charged, the neutron has no charge. Therefore, the nucleus is positively charged.

Evidence for the proton was discovered by Ernest Rutherford in 1919. The neutron was discovered by James Chadwick in 1931, but it was not until the mid-1950s that physicists found evidence of smaller particles inside the proton and neutron. These particles were named quarks.

There are two types of quark, up quarks and down quarks.

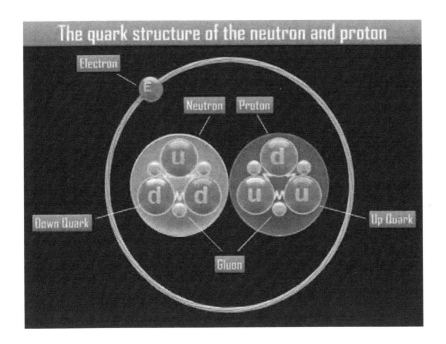

Figure 13[13]

The proton contains two positively charged up quarks and one negatively charged down quark. The neutron contains two negatively charged down quarks and one positively charged up quark. Protons and neutrons can no longer be regarded as fundamental particles. They are composite particles made up from fundamental quark particles.

---

[13] General-fmv/Shutterstock

There are just three fundamental matter particles involved in the atom's structure, and as a result, everything around us, because all the matter in the universe is made from electrons, up quarks and down quarks.

MATTER
from molecule to quark

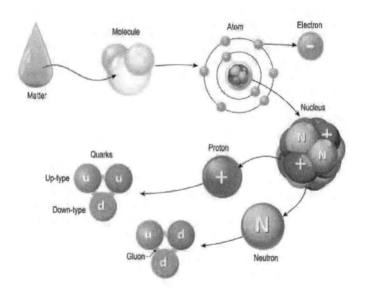

Figure 14[14]

---

[14] Designua/Shutterstock

Figure 14 shows the complete picture from matter to quarks. The three fundamental particles in the atom are from a group of particles called fermions. No two identical fermions can occupy the same quantum state in an atom. This is the essence of Pauli's rule, which we'll discuss in Chapter 6.

The forces of nature

There are four fundamental forces of nature, carried by particles called bosons:

- Gravity, as discovered by Isaac Newton in the 17th century
- The electromagnetic force was first introduced by James Maxwell in1864
- The strong nuclear force was discovered by James Chadwick in 1921
- The weak force involved in decay, first proposed by Enrico Fermi in 1933

As you will see in Figure 15, the force of gravity is not included in the standard model. Indeed, according to Einstein it is not a force at all. We have already discussed how weak gravity is at the human scale. It is negligible at the atomic scale

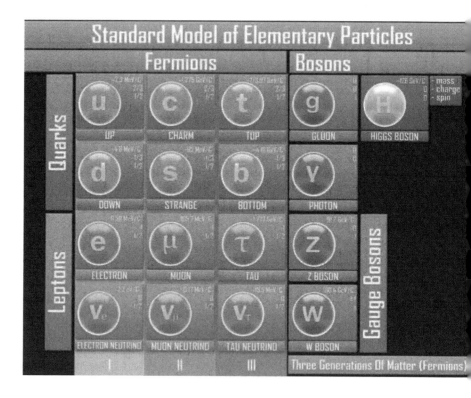

Figure 15[15]

This standard model of elementary particles shows all the particles known to date. Don't be too concerned with the names on the model, they are just labels invented by physicists with a sense of humour. For the purpose of this book, we will only need the particles directly involved in the

[15] general-fmv/Shutterstock

atom's structure, so please ignore the particles in columns II and III. These are identical to the atomic particles in column I, but bigger. It is not yet known what they do; it's possible that they are clues to things still undiscovered. Only time will tell.

As we have seen, the particles involved in the atom are:

- Quarks – the first two in column I, marked u and d on the chart.
- The electron – the third down in column I of the chart and marked e. These particles are in the group known as fermions.

The three forces of nature involved in the atom are carried by the four particles listed under bosons on the chart. The g is the gluon – the carrier of the strong nuclear force. As the name implies, this is the strongest force in nature and binds the particles of the nucleus together.

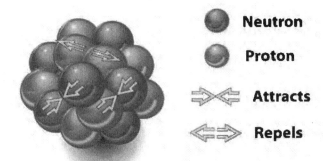

Neutron

Proton

Attracts

Repels

Atomic Nuclei and the Strong Force

Figure 16

16

The photon y is the carrier of the electromagnetic force which acts between the positive nucleus and the negative electrons which orbit the nucleus. The z and w bosons are involved in the weak nuclear force.

The weak nuclear force differs from the other forces in that, rather than holding things together, it breaks things apart and is involved in decay. This force works over very short distances.

In 1964, François Englert and Peter Higgs separately theorised the Higgs boson H, but it was not until 2013 that physicists at the LHC in Switzerland announced that they had finally found it. The Higgs boson carries the Higgs

---

[16] OsweetNature/Shutterstock

field, which permeates space and gives atoms of different sizes their mass.

The neutrino and electron neutrino

That leaves just one particle unexplained: the electron neutrino (bottom of column I, marked Ve).

Neutrinos are emitted when the Sun converts hydrogen into helium to fuel itself. This process is known as nuclear fusion, explained further in Chapter 7. Neutrinos are similar to electrons, but have no electric charge and only a tiny mass. They do not interact with matter and are emitted in huge amounts – 65 billion per second are travelling through every square centimetre of matter. We can't escape them at night, even though the Sun is on the other side of the planet; they pass through everything, including you, me and the Earth, as if it wasn't there. Huge detectors have been built deep underground to facilitate ongoing research.

There are three types of neutrino corresponding to three types of electron in the standard model. All are on the bottom line of the chart: electron, muon and tau. As we are ignoring the two middle columns for the purposes of this book, we are left with the electron neutrino, Ve.

The electron neutrino is involved in beta decay in atoms. When there are too many protons or neutrons in an atom's nucleus, beta decay occurs.

Beta decay is the process in which atoms release beta particles in order to conserve energy. There are two types of beta decay, positive and negative. Positive beta decay releases a positively charged positron and a neutrino. By far the more common type of beta decay is negative. This decay releases an electron and an antineutrino.

Electrons are affected by the electromagnetic force. Neutrinos are only affected by the weak nuclear force which operates over very short distances, enabling them to pass through matter.

The positron and antineutrino particles are examples of antimatter.

Antimatter

Antimatter is not, as some may think, fictional. It is not a myth from the worlds of *Star Wars* or popular stories. Antimatter is very real.

It was first proposed by Paul Dirac in 1928. Then in 1932, Carl Anderson discovered an antiparticle which was the opposite in every way to an electron.

He named this particle a *positron*.

One of the differences is its polarity. As you know, the electron has a negative charge, whereas the positron has a positive charge. Hence the name.

When matter and antimatter particles collide, they instantly annihilate each other and energy is produced. Physicists now know that every particle in the standard model has an antimatter equivalent.

We will cover the Big Bang, the birth of the universe, in more detail in Chapter 7. For now, we'll just touch on it.

In the first second after the Big Bang, all of the fundamental particles, from which all matter is made, came into existence. At the same time, all antimatter particles were born. Both types of particles had the same mass, but all other characteristics, such as charge, were opposite. It is believed that there were equal amounts of matter and antimatter at this stage.

As the universe cooled and expanded, most matter and antimatter particles annihilated each other. Of those that were left, slightly more were matter than antimatter. Some physicists believe that neutrinos may have been involved in that process, and studying neutrinos may explain why the universe is made up of matter rather than antimatter.

The imbalance at the start has become vastly pronounced today. It is thought that had there not been that initial imbalance, matter and antimatter would have annihilated each other by now and the universe would be filled with nothing but photons.

Matter and Antimatter Atoms

+ ● Proton
o ● Neutron
– • Electron

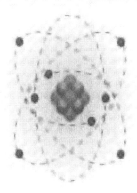

– ● Antiproton
o ● Antineutron
+ ● Positron

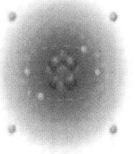

Figure 17[17]

The Periodic Table of Elements

We will not be going into the Periodic Table in great detail.
All we need here is knowledge of how atoms are related to
it.

---

[17] chromatos/Shutterstock

A chemical element is a substance in which all atoms have the same number of protons in their atomic nuclei. The number of protons in an atom's nucleus defines its atomic number in the Periodic Table of Elements, and thus defines the element of which it is a component part (see Fig18). For example, the atomic number of oxygen denotes that each of oxygen's atoms has eight protons. The total number of protons and neutrons in an atom is its atomic mass.

A neutral atom has the same number of protons with positive charge and electrons with negative charge. It also has the same number of protons as neutrons. If there are fewer electrons than protons in an atom, it is an ion and will have a net positive charge. If there are fewer neutrons than protons in an atom, it is an isotope and its atomic mass will be less than the neutral atom of the same element. Therefore, ions and isotopes are different versions of the same element.

But we don't need to know any more about ions and isotopes here. We are in danger of straying into chemistry.

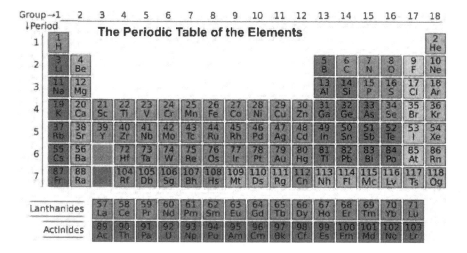

Figure 18[18]

Let's look at hydrogen, symbol H, in the top left-hand corner of the Periodic Table. The number above the H is this atom's atomic number, in this case 1. We know that this is the number of protons in this element's atom, and in a neutral atom there are the same number of electrons as protons.

Hydrogen is the smallest atom and differs from other atoms as it has no neutron. The proton and electron balance out to neutral and it has a tiny mass. Hydrogen is the most abundant element in the universe.

---

[18] cosept w /Shutterstock

Figure 19: Hydrogen atom[19]

Let's look at another example, oxygen, symbol O, towards the right on the chart. We can see that oxygen's atomic number is 8, which means it has eight protons, and being a neutral atom, it will also have eight electrons and eight neutrons.

[19] general-fmv/Shutterstock

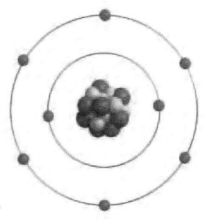

Figure 20: Oxygen[20]

As we can see from Figure 20, oxygen's electrons are arranged in energy levels. There are a maximum of two in ground state (the inner shell) and six in the next shell, which can hold a maximum of eight (see Fig 11). There are no other shells (energy levels) in this case.

The water molecule is H2O, i.e. two atoms of hydrogen connected to one atom of oxygen. Oxygen has six electrons in the second energy level, so that level has two spare places. Two hydrogen atoms with only one electron each can attach themselves to the spare places in the oxygen shell and we have a water molecule.

See figure 21.

[20] ooaka/Shutterstock

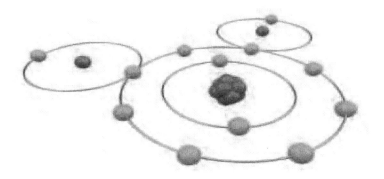

Figure 21[21]

Let's have a look at some more examples. Iron, symbol Fe (middle of top row), has an atomic number of 26. In this neutral atom, there are twenty-six neutrons and twenty-six electrons arranged as in figure 22.

---

[21] IQoncept/Shutterstock

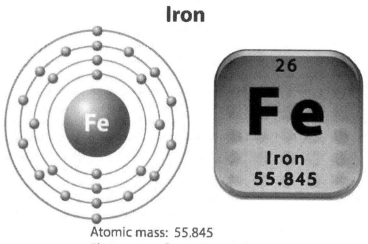

Iron

Atomic mass: 55.845
Electron configuration: 2, 8, 14, 2

Figure 22[22]

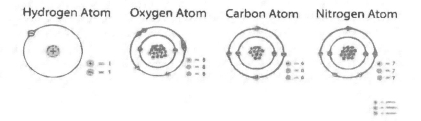

Drawing the Simplest Atoms

Hydrogen Atom   Oxygen Atom   Carbon Atom   Nitrogen Atom

Figure 23[23]

[22] BlueRingMedia/Shutterstock

[23] Naski/Shutterstock

Chapter 6 – Quantum Matters

In 1900, German physicist Max Planck demonstrated that radiant energy is made up of particle-like components, quanta, and particles can have wavelike characteristics. Quantum physics was born and Max Planck is widely regarded as its father.

In this chapter, we will cover:

- Young's double slit experiment
- Heisenberg's uncertainty principle
- The electron cloud model
- The heart of quantum mechanics
- Pauli's exclusion principle
- Quantum fields

One very important concept in quantum theory is Pauli's exclusion principle. Another is Heisenberg's uncertainty principle. But we'll begin by exploring the famous double-slit experiment.

Young's double slit experiment

In the 17th century, Isaac Newton concluded that light is carried by corpuscles (particles). Christiaan Huygens argued that Newton was wrong and light is a wave, but Newton's theory was accepted because of his greater prestige.

In 1801, long before quantum physics was even thought of, Thomas Young performed his famous double-slit experiment showing wave diffraction and interference, proving that light is a wave. The double-slit experiment requires a monochromatic light source, i.e. a single wavelength of light. It is unclear how Young achieved this as, unfortunately, he never recorded his process.

Different frequencies of the light spectrum make different interference patterns. All the frequencies together (white light) would produce a very blurry composite of all the patterns in this experiment.

Diffraction of waves can be observed when water flows through a gap or around an obstacle. You see diffraction as the water bends around the edges of the gap or obstacle, producing waves as it emerges. Light performs in the same way.

In Figure 24, when the light passes through the double slits in the barrier, we see two distinct waves emerging. When the two waves emerge, they interfere with each other in two different ways, as shown in Figure 25.

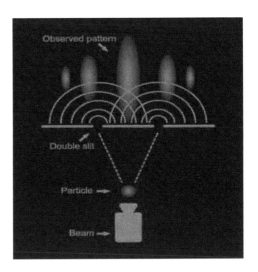

Figure 24[24]

[24] Photos by D/Shutterstock

Wave interference

Figure 25[25]

Where the crest of a wave meets with the trough of the other, they are reinforced. This is constructive interference. Where the waves meet crest to crest, they cancel each other out. This is destructive interference.

Where there is constructive interference in the light wave, the screen at the back shows bright stripes. When there is destructive interference, there are dark stripes on the screen (see Fig 24). This indicates that light is a wave.

You may conclude from this that Huygens should have won the argument with Newton in the 17th century. But read on.

In 1905, Albert Einstein, in his work on the photoelectric effect, proposed a quantum of light (the photon) which behaves like a particle and a wave. In other words, light is

---

[25] Naski/Shutterstock

both particle and wave. This can be demonstrated using another version of the double-slit experiment (Fig 26).

Diffraction and Interference

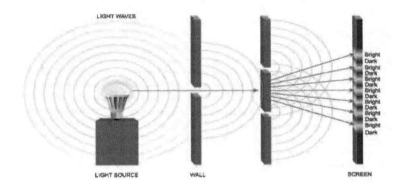

Figure 26[26]

When monochromatic light passes through the single slit in the first barrier, it diffracts and emerges as a wave, but appears to go through to the barrier behind in a straight line as a particle.

If we remove the first barrier and send light through both slits, the interference pattern of waves appears. If we now

---

[26] Designua/Shutterstock

72

close one slit and send light through, it behaves as a particle as before. Einstein was right – light is both wave and particle.

In 1927, Clinton Davisson and Lester Germer showed experimentally that electrons perform like waves. Their experiment showed a diffraction pattern when electrons were scattered by the surface of a crystal of nickel metal. This supported Louis de Broglie's hypothesis of 1924 in which he postulated that matter had wavelike properties.

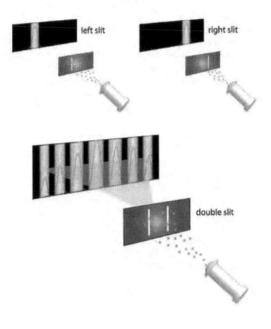

Figure27[27]

The double-slit experiment was first performed using a
stream of electrons by Claus Jönsson in 1961. Diffraction
and interference were displayed just as for light, indicating
that electrons could be both particle and wave (see Fig 27).
The double-slit experiment can now be performed using
neutrons, atoms and some molecules.

---

[27] Maagnitix/Shutterstock

Antimatter was shown to behave in exactly the same way as matter by physicists in 2018. They developed a double-slit experiment using positrons (antiparticles of electrons). The double-slit experiment can also be used to demonstrate how uncertainty and probability are prevalent in any quantum system, as we will soon discover.

Heisenberg's uncertainty principle

The uncertainty principle, introduced during the mid-1920s by Werner Heisenberg, states that:

Both the location and the momentum of a particle cannot be known simultaneously.

We can know the path of an electron as it moves through space or we can know where it is at a given position. But we cannot know both. If we observe where electrons are, we cannot know their momentum and vice versa. We can only state probabilities of where particles may be or what their momentum is. This sets a limit to what we can predict in quantum physics.

The electron cloud model

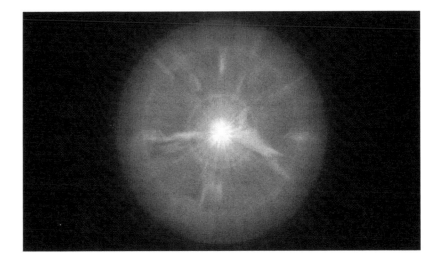

Figure 28 (artist's impression)[28]

The electron cloud model was introduced by Erwin Schrödinger in the mid-1920s. This current model of the atom predicts clouds of probability of where the electrons are around the nucleus. We can only say what regions the electrons are likely to be in. Pauli's exclusion principle, discussed later in this chapter, still applies. The exclusion principle is easier to visualise using Bohr's atomic model.

Let's take another look at the double-slit experiment to see how observing (measuring) an electron affects it.

---

[28] Magnetix/Shutterstock

The heart of quantum mechanics

The double-slit experiment can now be performed to show the duality of matter and how observing the experiment can affect the result. The following was a thought experiment in 1978. It was performed using modern technology in 2007.

Figures 29 and 30 show the results of observing (measuring) an electron's position. *A device which fired electrons one at a time was used.* Firing electrons one at a time with both slits open results in the interference pattern appearing on the back screen, just as it did with a stream of electrons. Each single electron appears to go through both slits at the same time as a stretched wave (see fig 29).

Figure 29[29]

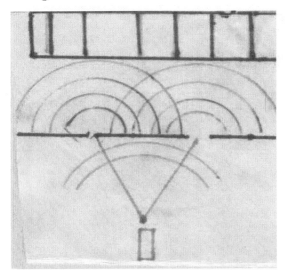

If we now place sensors to observe each electron passing through the slits (Figure 30) we find that 50% of the time the electrons go through the left slit and the other 50% of the time they go through the right slit. But the interference pattern does not show on the back screen. What we see are two stripes of electrons on the back screen directly behind each slit.

---

[29] Author's own image

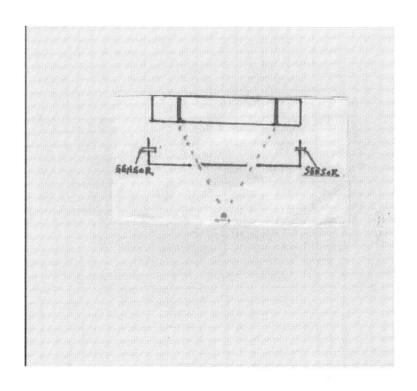

Figure 30[30]

The only difference in the experiment this time is the addition of sensors to observe the electrons. It's as if the electrons know they are being watched, because if we remove the sensors, the interference pattern will appear on the back screen again. Are you confused? If you are, you're in good company.

---

[30] Author's own image

Asking how that can happen demonstrates the measurement problem in quantum theory. Richard Feynman, one of the most brilliant physicists of the 20th century, is quoted as saying:

"We choose to examine a phenomenon which is impossible, absolutely impossible, to explain in any classical way, and which has in it the heart of quantum mechanics."

Pauli's exclusion principle

If atoms are almost all empty space, why can't we walk through walls? This question was answered by Wolfgang Pauli in 1925.

We now know that electrons, like all matter, can be particles or waves. The exclusion principle applies to all of the fermions in the standard model (Fig 15), but not to bosons.

Electrons have four quantum numbers:

- The principal quantum number
- The orbital angular momentum quantum number

- The magnetic quantum number
- The electron spin quantum number

The exclusion principle states that no two electrons in an atom can have exactly the same four quantum numbers.

In Chapter 5, we looked at different models of the atom. We learned from Bohr's model that electrons orbit the nucleus in fixed shells. And we learned that all elementary particles of the same type are identical to each other and in pristine condition. But their quantum numbers differ.

Atoms differ from each other not because they have different types of electrons, but by the number of electrons they have and how they are arranged. An atom which makes up a solid will never allow another electron to join it.

Having said that, electrons will share with others if there is space. Recall from Chapter 5 how water (H2O) molecules are built (see Fig 21). Atoms in air (gas) and liquid are not as densely packed together as those in solids, so we can walk through them. Put your hand out through an open car window when the car is moving. The wind you feel is the atoms moving apart.

Let's look at the atom of iron.

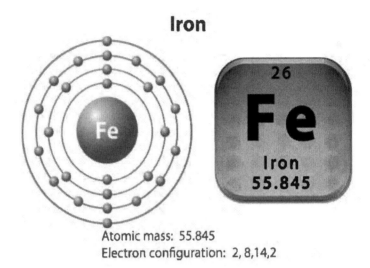

Atomic mass: 55.845
Electron configuration: 2, 8, 14, 2

Figure 31[31]

Pauli's exclusion principle prevents unwanted electrons entering an already made atom such as iron. But how?

If electrons were little round balls, it would be difficult to see how. But as we know that they can also be waves, then how they exclude unwanted electrons from an already made atom becomes easier to see.

Electronic waves can stretch to great distances, but can never overlap. The outer (or valence) shell in the iron atom

---

[31] Banderlog/Shutterstock

has two electrons. We can think of the stretched waves between electrons as barriers to keep unwanted electrons out by preventing them from overlapping. That is why the atoms in your feet could not force themselves through an iron floor you were standing on, or any other solid.

Quantum fields

We will not be looking too closely into quantum field theory (QFT) in this book, but I recommend more exploration in your future reading.

Recall the forces of nature we discussed in Chapter 5: gravity, electromagnetic, strong nuclear and weak nuclear, along with Higgs. They are carried by particles called bosons. The bosons operate in force fields which permeate all "empty" space everywhere in the universe. Indeed, this is true of all fundamental particles. In this sense, there is no empty space anywhere in the universe, but everything in the universe is connected. We and everything are immersed in a sea of invisible fields.

Chapter 7 – Cosmology

In this chapter, we will cover a lot, including:

- The Big Bang
- The early universe
- Stars and the evolution of bigger atoms
- Hubble's law and Hubble's constant
- Dark energy
- How inflation developed
- The big crunch
- The big freeze
- Dark matter
- The cosmic microwave background

The Big Bang

The Big Bang theory was first proposed by a Catholic priest, Georges Lemaître, in 1927. He deduced from Einstein's equations of general relativity that the universe expanded explosively from an extremely dense and hot point. It has since been estimated that it occurred 13.7 billion years ago.

Figure 32[32]

In the first seconds after the Big Bang, the fledgling universe was intensely hot and dense. It rapidly (in millionths of a second) cooled just enough for the four fundamental forces of nature to emerge: electromagnetic, weak nuclear, strong nuclear and gravity. The particles which are the building blocks of the whole universe also formed.

These particles are known today as electrons and quarks.

---

[32] VectorMine/Shutterstock

A few millionths of a second later, quarks joined to form protons and neutrons, and a few minutes later, protons and neutrons joined to form atomic nuclei.

The early universe

Around 380,000 years later, the electrons were trapped in orbit around nuclei and the first atoms were born: mainly hydrogen and helium, with traces of lithium and beryllium. These first elements were formed by thermonuclear fusion, the universe still being hot and dense enough for that process to happen. As the universe cooled and became less dense, fusion could no longer happen.

In the early universe, hydrogen and helium accounted for almost 100% of all the normal matter. Today, 13.7 billion years later, these two elements still account for 98% of normal matter in the known universe, but here on Earth, they are not so abundant. But without hydrogen, we would have no water (H20).

Hydrogen is the smallest atom, having only one proton and one electron. Helium is the second smallest with two protons, two neutrons and two electrons. So if these two atoms are 98% of the ordinary matter in the universe, where did the other larger atoms come from?

Stars and the evolution of bigger atoms

Stars are born. The universe was 1.6 to 3 million years old when gravity began to form stars from clouds of gas. That gas was awash with hydrogen and helium atoms.

Stars have their lives. The central core of a star, like our Sun, is very dense and hot (the temperature at the core of a star is sixteen million degrees Celsius), and with stars' massive gravity providing ample energy, the process of fusion can begin again. Electrons are ripped from their hydrogen atoms' nuclei.

The pressure at the core is immense because of the star's size and gravity, so the hydrogen nuclei are packed closely together and begin to fuse with each other, forming larger nuclei with varying numbers of protons and neutrons up to the size of iron. These large nuclei then move out of the core, and the electrons left behind by the hydrogen atoms are trapped in orbits around them and produce heavier atoms, including the atoms we are made from (we really are stardust). Successive generations of stars have been producing them ever since.

This process of making big atoms out of smaller ones is called thermonuclear fusion. Thermonuclear fusion should not be confused with thermonuclear fission. They are both physical processes that produce energy from atoms, but:

- Fusion joins two or more lighter atoms into a larger one
- Fission splits two or more large, unstable atoms into two or more smaller ones

Fusion takes place in the Sun, producing a huge amount of energy. If we were ever able to use it on Earth, all our energy problems would be solved. Fission is used in nuclear power stations to produce electricity.

Stars die. When a star's atomic fuel runs out, the fusion in the core stops and will no longer produce nuclei. The star shrinks and becomes more dense and fusion begins again, but outside the core. This renewed reaction overpowers gravity and the star's surface is pushed outwards.

Atoms larger than iron are produced (gold, lead, etc.). This reaction then overpowers gravity and the star grows and explodes, and all of the atoms it has made during its life are scattered into space.

Stars are reborn. From the debris of exploded stars, huge clouds of gases and atoms form. Gravity then squeezes these clouds together to form new stars and the whole process begins again. Physicists have deduced that our Sun is a third generation star and has another five billion years to live.

Hubble's law and Hubble's constant

The Big Bang theory was strengthened in 1929 when Edwin Hubble found that other galaxies are moving away from us, and the further they are from us, the faster they are moving. This is known as Hubble's law. Hubble used a technique known as redshift in his calculations. Redshift is similar to the doppler effect in sound waves.

You can hear an example of the doppler effect when a police car is coming towards you with its siren on. As it approaches, the siren is high pitched. After it has passed, the sound is lower pitched. This is caused by a change in pitch of the sound waves.

The same thing happens with light waves, but on a much finer scale. For far-away galaxies, the light waves will be stretched towards the red end of the light spectrum. This occurs not simply because of the distance, but also because of the galaxies' acceleration.

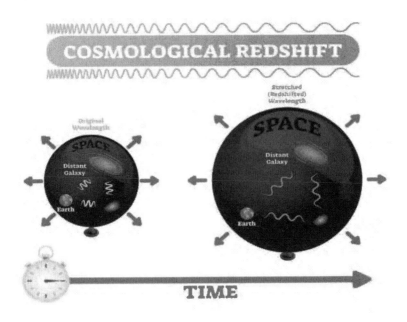

Figure 33[33]

---

[33] Designia/Shutterstock

A good analogy for the expanding universe is to imagine a plum pudding baking in an oven. As the mixture cooks, it expands. Imagine that you are sitting on a plum in the mixture. The plums closest to you would be moving away more slowly than the ones further away as the mixture expands. Spacetime is expanding in the same way as the mixture and has been doing so for 13.7 billion years.

Note that the galaxies (and the plums) are not moving away *through* spacetime. It is the spacetime which is expanding in a similar way to the air in an inflating balloon.

Figure 34[34]

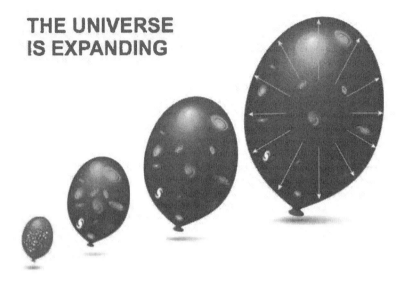

THE UNIVERSE IS EXPANDING

The speed of the expansion of the universe (but thankfully not the pudding) is given by Hubble's constant, which states that:

The speed at which galaxies move away from each other as spacetime stretches is 41.8 miles (67.4 kilometres) per second for every 3.3 million light years.

3.3 million light years = 1 megaparsec. Megaparsecs are a way of measuring distance in intergalactic space.

---

[34] VectorMine/Shutterstock

For example, for the first 3.3 million light years (1 megaparsec), spacetime is expanding at 41.8 miles (67.4 kilometres) per second. For the next megaparsec, spacetime is expanding at 96 miles (134.8 kilometres) per second. For the third megaparsec, spacetime is expanding at 144 miles per second (3 megaparsecs), and so on *to a very high speed where spacetime is expanding faster than the speed of light.* Spacetime is not only expanding, it is accelerating.

There has been much discussion regarding the correct value for the expansion of the universe and its value has changed several times. The figures I've given above are the current value, but this may change again

The acceleration is fuelled by dark energy.

Dark energy

In 1998, three physicists, Saul Perlmutter, Brian Schmidt and Adam Reiss, agreed after studying supernovae that five billion years ago, the universe's expansion began to accelerate. Until then, physicists had thought that the expanding universe should be slowing, not increasing.

After this revelation, physicists realised there must be an anti-gravitational force which repels rather than attracts matter. Calculations showed that there must be a vast amount of it. This force is now known as dark energy. It is believed that the universe consists of 68.3% dark energy (see Fig 35).

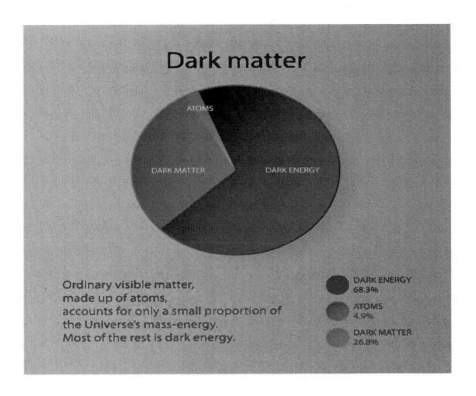

Figure 35[35]

---

[35] Naski/Shutterstock

How inflation developed

After the Big Bang, there was a massive rate of inflation. In the first second, the universe grew to the equivalent of just over one light year, then slowed. After one year, it had grown as large as our current Milky Way galaxy: 105.7 light years in diameter. Five billion years ago, the acceleration increased again under the influence of dark energy.

Because the universe is expanding and the distant objects we see now have been carried by the expansion and acceleration of spacetime to 46.5 billion light years away, the light we are seeing is not from where the galaxies are now, but from where they were when they were closer to us. The light is only just reaching us. It has been estimated by modern physicists that the whole of the universe, the light of which we will never see, is 250 times larger than the observable universe.

The biggest problem for our understanding of the universe is its size. The human brain finds it difficult to come to terms with how vast it is.

Because the universe is expanding, we know that it must have been smaller in the past. This supports the Big Bang theory. But how will it develop in the future? Will the universe grow to infinity or will it end sometime? If so, how will it end?

Several theories have been suggested to answer these questions. We will take a very brief look at two of them here.

The big crunch

The universe is finely balanced between the repulsive force of dark energy and the attractive force of gravity. If dark energy loses the battle with gravity, at some point in the future, the universe could be crunched back to its beginning.

There could be a "big rebound" and the whole thing would start again. This may already have been happening and our universe is just another generation of the process.

The big freeze

Alternatively, if dark energy wins the battle, the universe could keep expanding and accelerating until eventually the heat disperses throughout the expansion and all matter – galaxies, stars, gas clouds – would be driven further apart. Stars would no longer be able to form. All existing galaxies would eventually die and the whole of the much larger universe would be cold, dark and empty.

Dark matter

In 1933, Jan Oort calculated that the velocities of stars orbiting galaxies are far too high for the total mass of the galaxy. He estimated how much matter was contained in our galaxy, the Milky Way. Then he measured the velocities of the stars. He calculated that there should be five times more matter than is visible.

The extra mass is what we now call dark matter.

Even though we can't see dark matter, we know it's there because of its gravity. It bends light around itself just as normal matter does and can be detected by gravitational lensing, even though it appears that there is nothing there to bend the light.

We know it is not normal matter. We know it's not antimatter. We know it's there, but we don't know what it is yet.

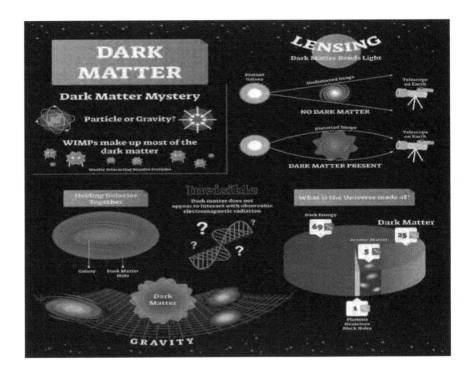

Figure 36[36]

We do know that the universe consists of 26.8% dark matter and 4.9% ordinary matter, plus 68.3% dark energy. Weakly interacting massive particles (WIMPs) are hypothetical particles. There is no evidence for them.

---

[36] trekandshoot/Shutterstock

The cosmic microwave background

More evidence for the Big Bang theory emerged in 1965 when two physicists, Arno Penzias and Robert Wilson, were testing some new apparatus which was designed to detect low-level microwave radiation. The apparatus was receiving interference similar to that which you see on a TV screen when the set is not tuned to any channel (see Fig 37).

Figure 37[37]

After trying various remedies, they concluded the interference was not caused by their apparatus, but must be

[37] nicemonkey/Shutterstock

coming from space, and it was coming from every direction. Shortly after this realisation, they became aware that another group of physicists had predicted that there should be residual cosmic microwave background (CMB) radiation left over from the Big Bang. While that group was developing equipment to detect it, Penzias and Wilson had found the evidence of CMB by accident. They received the Nobel prize for their breakthrough in 1978.

So, what is the CMB?

Recall that in the first seconds of the Big Bang, quarks and electrons came into existence. Milliseconds later, the quarks joined together to form protons and neutrons. Then they too joined to form nuclei. The expanding universe was filled with a thick, very hot and dense hydrogen plasma.

Light (photons) could not escape this opaque plasma until 380,000 years later when electrons became trapped in orbit around nuclei, and hydrogen and helium atoms were formed. Now photons could flow easily through space rather than being scattered by electrons in the fog.

Those photons gave the universe its first light and are what we now see in the CMB (Fig 38). It can be detected in every direction and it is *almost* the same in every direction.

Figure 38[38]

Almost, but not quite the same. The CMB shows some very subtle variations in its signal. The hot (lighter) and cold (darker) spots show temperature differences of only millionths of a degree and tiny density differences in matter collecting together in the early Universe. Over billions of years, gravity drew the denser spots together. They now correspond to the galaxy clusters that we see today.

The CMB, Hubble's law and Hubble's constant provide convincing evidence for the Big Bang theory. There are

---

[38] NASA/Wiki commons

various discussions amongst theorists about possible "other" universes. Examples of these are:

- Multiverses (a group of multiple universes)
- Infinite universe (an infinite number of galaxies in one universe.
- Parallel universes (a self-contained plane of existence, co-existing with our own)

There is no evidence for these theories yet and, fascinating though they are, they're beyond the remit of this book.

Conclusion

If a tree falls in the forest and there's no one there to hear it, will it still make a sound? If we humans are the only creatures in the universe with consciousness, would the universe still exist if we were not here to observe it? (Given the huge numbers of stars and planets in the universe, it is very unlikely that we are alone.)

These are philosophical rather than physics questions, but sometimes these two subjects can be closely connected.

A theory introduced by Brandon Carter in the 1970s known as the anthropic principle states that the universe is as it is because if it were not, we would not be here to observe it. If any small differences had occurred at, or soon after, the Big Bang, the universe we know would not exist, and consequently neither would we. If any of the fundamental forces or particles changed even very slightly, it would be a completely different universe.

Who are we? Where are we? What are we? Why are we? What's going on? Reading popular science will not give definitive answers to these questions. Instead, it will raise more questions. It will exercise your brain, giving you hours of fascinating and absorbing entertainment. But it's important to enjoy it to whatever level you are comfortable with.

Question everything and enhance your understanding. Even Einstein was not always right. Neither am I.

If you enjoyed this book, or have any comments, please leave a review on the amazon website or elsewhere.

Endnotes

Image references:

Professor's Diamond	Lifetime stock/Shutterstock
Figure 2	Designua/Shutterstock
Figure 3	MicroOne/shutterstock
Figure 4	Fouad A Saad/Shutterstock
Figure 5	PRILL/Shuttersock
Figure 6	Rost9/Shutterstock
Figure 7	NASA/Wiki commons
Figure 8	NaracOrn/Shutterstock
Figure 9	Emir Kaan/Shutterstock
Figure 10	udaix/Shutterstock
Figure 11	Naski/Shutterstock
Figure 12	Naski/Shutterstock
Figure 13	general-fmv/Shutterstock
Figure 14	Designua's/Shutterstock
Figure 15	general-fmv/Shutterstock
Figure 16	OsweetNature/Shutterstock
Figure 17	chromatos/Shutterstock
Figure 18	cosept w /Shutterstock

Printed in Great Britain
by Amazon

55852531R00061